I dedicate this book to my daughter, Aubree. One day, we came across a rainbow and joyfully drove under it. We looked back and it was gone. The memory of this once in a lifetime event, is ours forever.

May we all cherish the magic in life.

Other Books by the Author

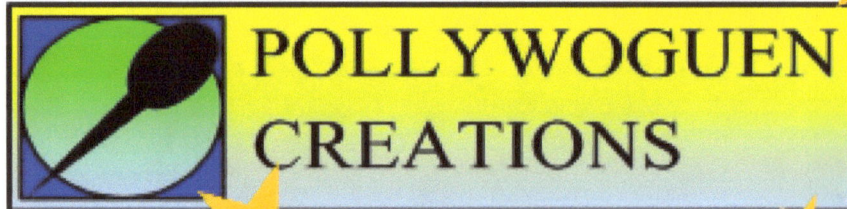

Walrus, Will You Hang With Me?

I Lost My Tooth

Where is Cleo?

POLLYWOGUEN CREATIONS

All About Rainbows

Written and illustrated by Robert Carr

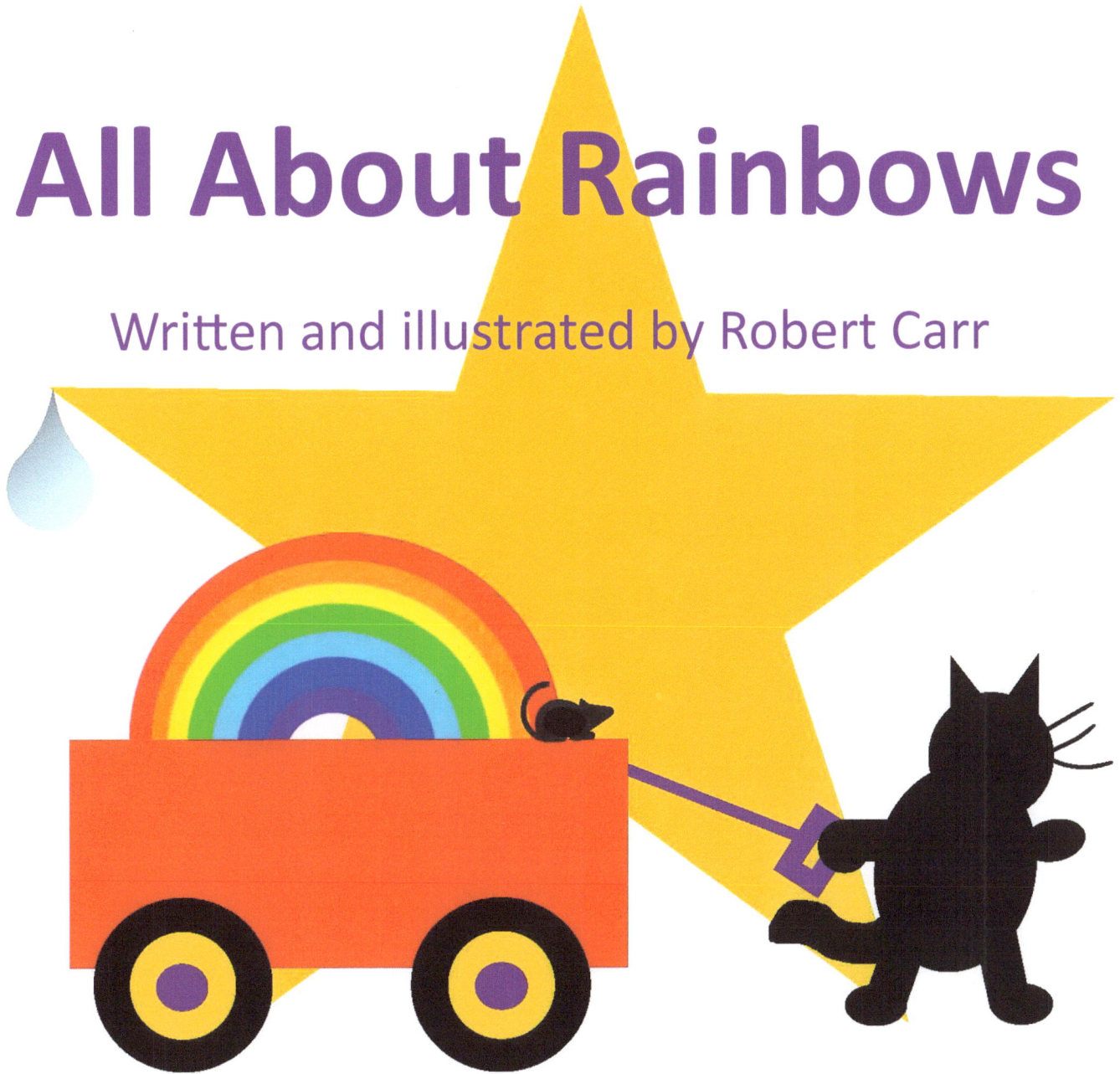

My name is Cleo. I am a cat.
I love rainbows. Do you?
If you do, then you and I will get
along just fine. Come into this book
and hear what I have to say. I will
help you love rainbows more than
you do now. Meow...

Red
Orange
Yellow
Green
Blue
Indigo
Violet

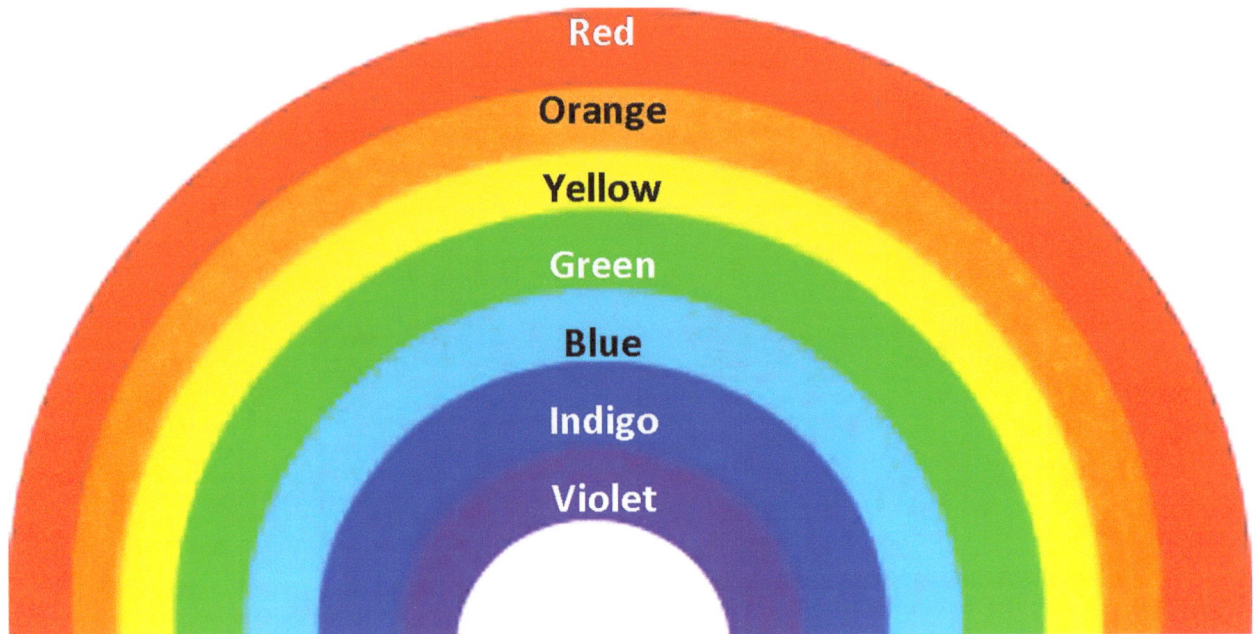

There are seven colors in the rainbow. The number of colors is easy to remember because the number seven appears in so many ways in your world and mine.

| 1 | 2 | 3 | 4 | 5 | 6 | 7 |

The Seven Dwarfs

The Seven Days of the Week

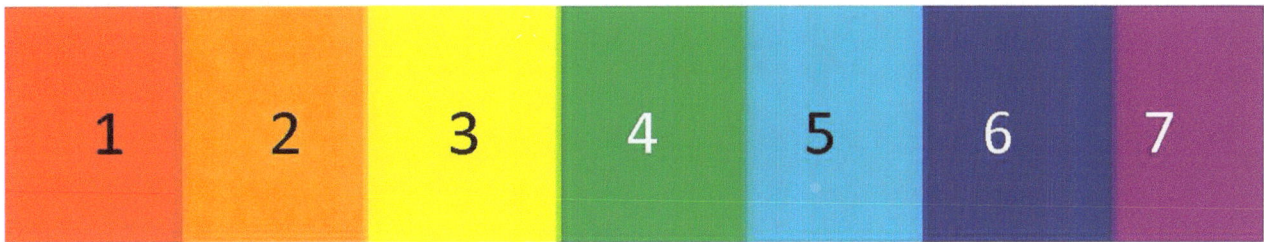

1	2	3	4	5	6	7
Sun.	Mon.	Tues.	Wed.	Thurs.	Fri.	Sat.

The Seven Notes on the Music Scale

C	D	E	F	G	A	B	C
Do	Re	Mi	Fa	So	La	Ti	Do

The Seven Continents

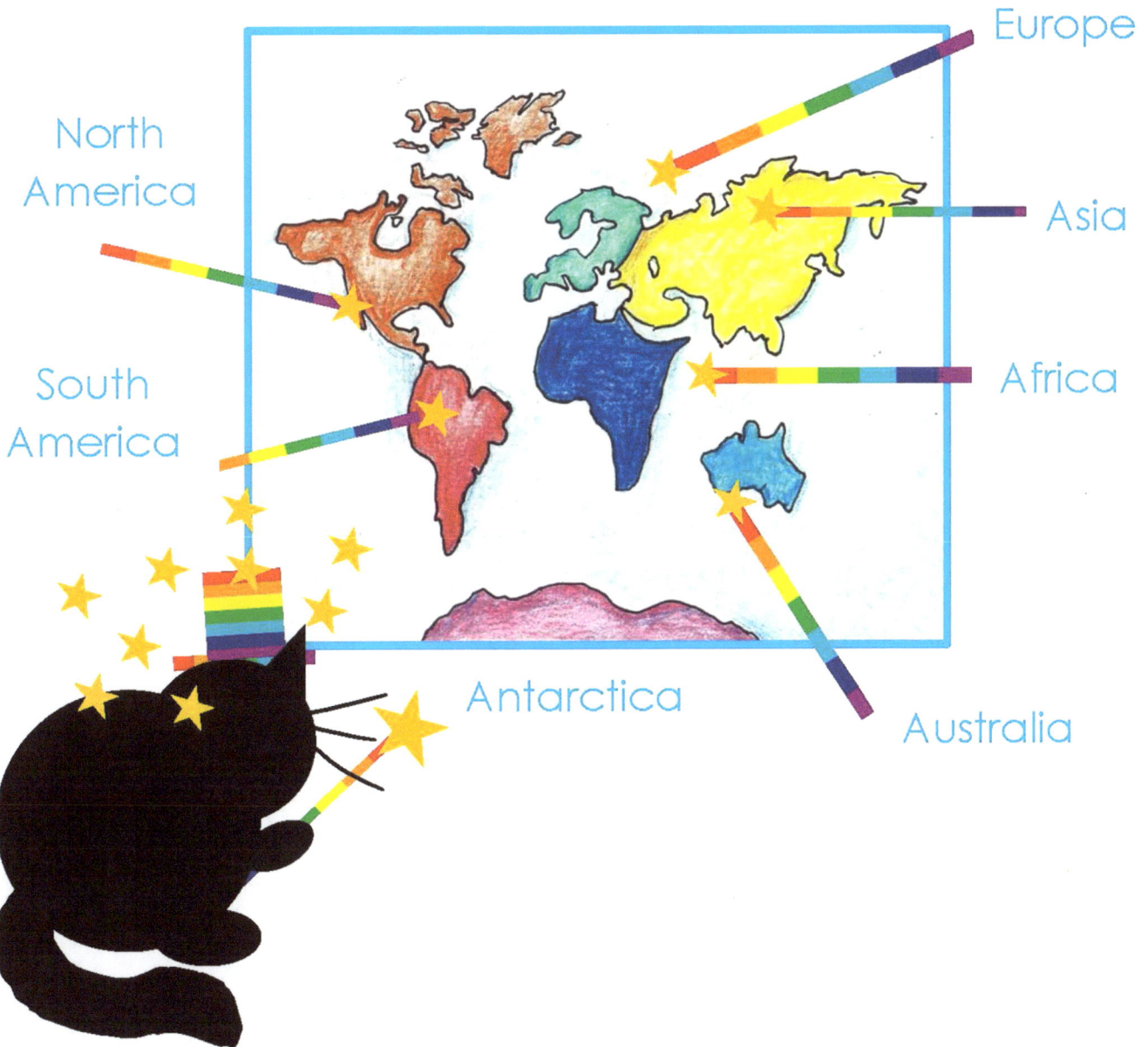

North America

Europe

Asia

South America

Africa

Antarctica

Australia

Seven Types of Rainbows

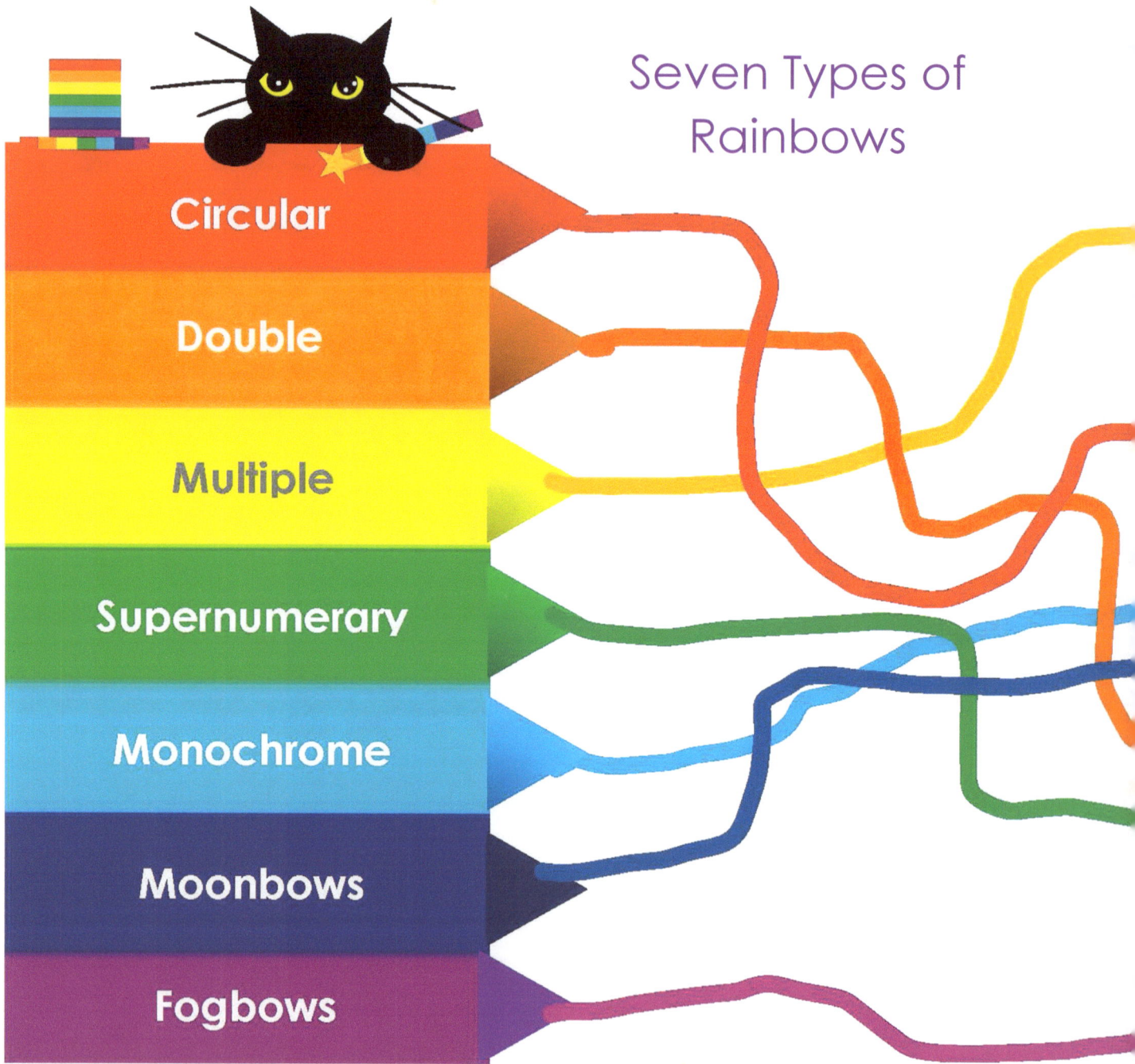

Circular

Double

Multiple

Supernumerary

Monochrome

Moonbows

Fogbows

Are when more than two rainbows form at the same time. One is strong and the others are faint in color.

Are the rainbows we see most often. They usually appear as a half circle because we are seeing them from the ground.

Are when one rainbow is above another. The top bow is lighter and the order of colors are in reverse. It occurs when rays of sunlight are reflected twice within raindrops.

Are rainbows that appear in the night when the moon is full. The light from the moon is refracted through water droplets in the air. Moonbows are not as bright as circular rainbows. They can be bow-shaped or a full circle.

Are when only one red bow appears. This happens when the sun is red and setting.

Are faint bow lines that appear just inside and under the main rainbow and usually echo the same color pattern.

Are soft bows of white in a cluster of fog. To find one, the sun must be behind you, the fog in front of you and low to the ground. The sky must be mostly clear.

I think it's time to be playful.
Rainbow is a colorful word. When I
say the word rainbow, sometimes I
see that it is raining bows.

Tickle my fancy. Oh, what fun!
How do you see raining bows?

Do you know how you can remember the order of the colors in a rainbow? I have a friend who can help you do so. His name is **Roy G. Biv**. Here he is! Take a very close look.

R
O
Y
G.
B
I
V

Red
Orange
Yellow
Green
Blue
Indigo
Violet

Roy G. Biv

I have the recipe to make a real rainbow. We just need three things: **Rain**, **sun**-with its light rays, and our **eyes**.

To see a rainbow, you will have to look away from the sun like I am doing. You must look up as well.

When I see a rainbow, it is usually in the shape of a bow or a half circle.
But rainbows are really completely round. From the ground, we can only see half of the circle because of where we are standing and looking from. This is called our point of view.

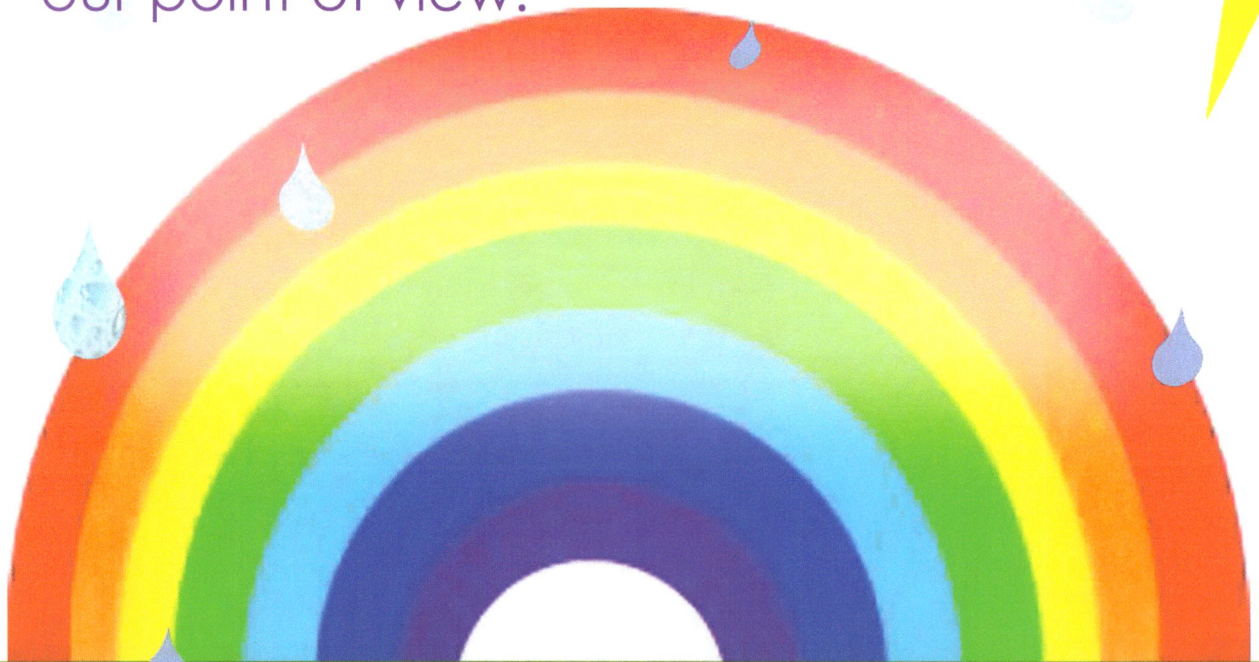

If we are looking from a high mountain or from an airplane, we may be able to see a rainbow as it really is. It would be a full circle. Therefore, there is no real end to a rainbow.

Did you know
you cannot
touch a real
rainbow?

To touch a rainbow would be like
trying to pick up your own shadow.
A rainbow is an illusion. Yet, our hearts
can be touched by a rainbow.

A rainbow's shape is called a pessimistic shape because it shaped like our mouth when we frown. An optimistic shape is one shaped like a smile. I am puzzled. Does a rainbow's shape look sad?

A clown with a rainbow frown,
A clown with a rainbow smile,
Neither rainbows will let you down.
Their magic will last a while.

Love Cleo

Seven Wise Thinkers

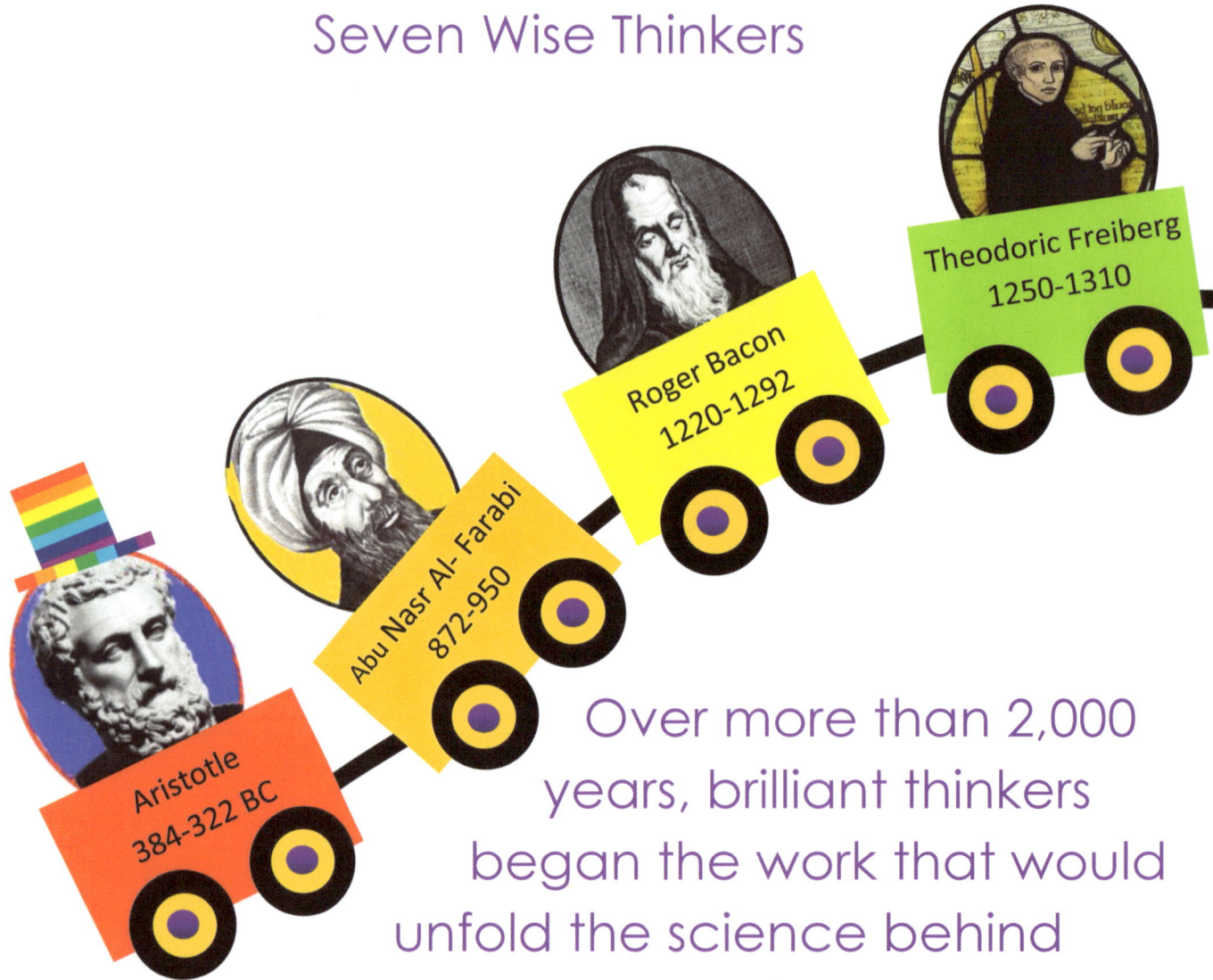

Theodoric Freiberg
1250-1310

Roger Bacon
1220-1292

Abu Nasr Al- Farabi
872-950

Aristotle
384-322 BC

Over more than 2,000 years, brilliant thinkers began the work that would unfold the science behind what creates a rainbow.

Smart eggs, I say.

Rene Descartes
1596-1650

Isaac Newton
1642-1726

Thomas Young
1773-1829

It seems to me, that curiosity brings us to many places.

I think I will learn how to fly. What do you want to learn?

Learning seems magical to me. The good news is, we can all learn something new seven days a week. Awesome!

The scientists of today believe dinosaurs of the distant past could see colors like you and me.

If that is true, it is possible that dinosaurs did see magical rainbows too.

It is likely that rainbows are as old as the Earth. As long as there has been sun and water, there has been the possibility of rainbows.

It is no wonder that over so many years the peoples of the Earth made up many stories about rainbows.

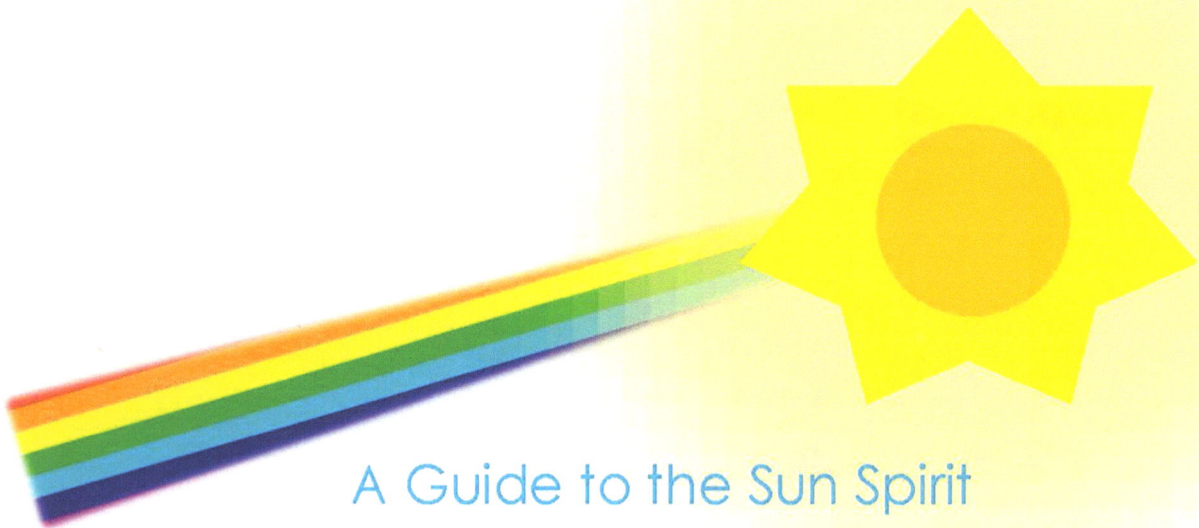

A Guide to the Sun Spirit

(Siberia)

A Pathway
To a Pot of Gold
(Ireland)

Goddess Iris Appears as a Rainbow
(Greece)

Rainbow Serpent as a Creator of Life
(Australia)

Nut - Goddess of the Sky
(Egypt)

Sun God with Rainbow Hem

(Native American - Cherokee)

Would you like to make a rainbow?
We will need a sunny day and a garden
hose with a spray nozzle attached to it.

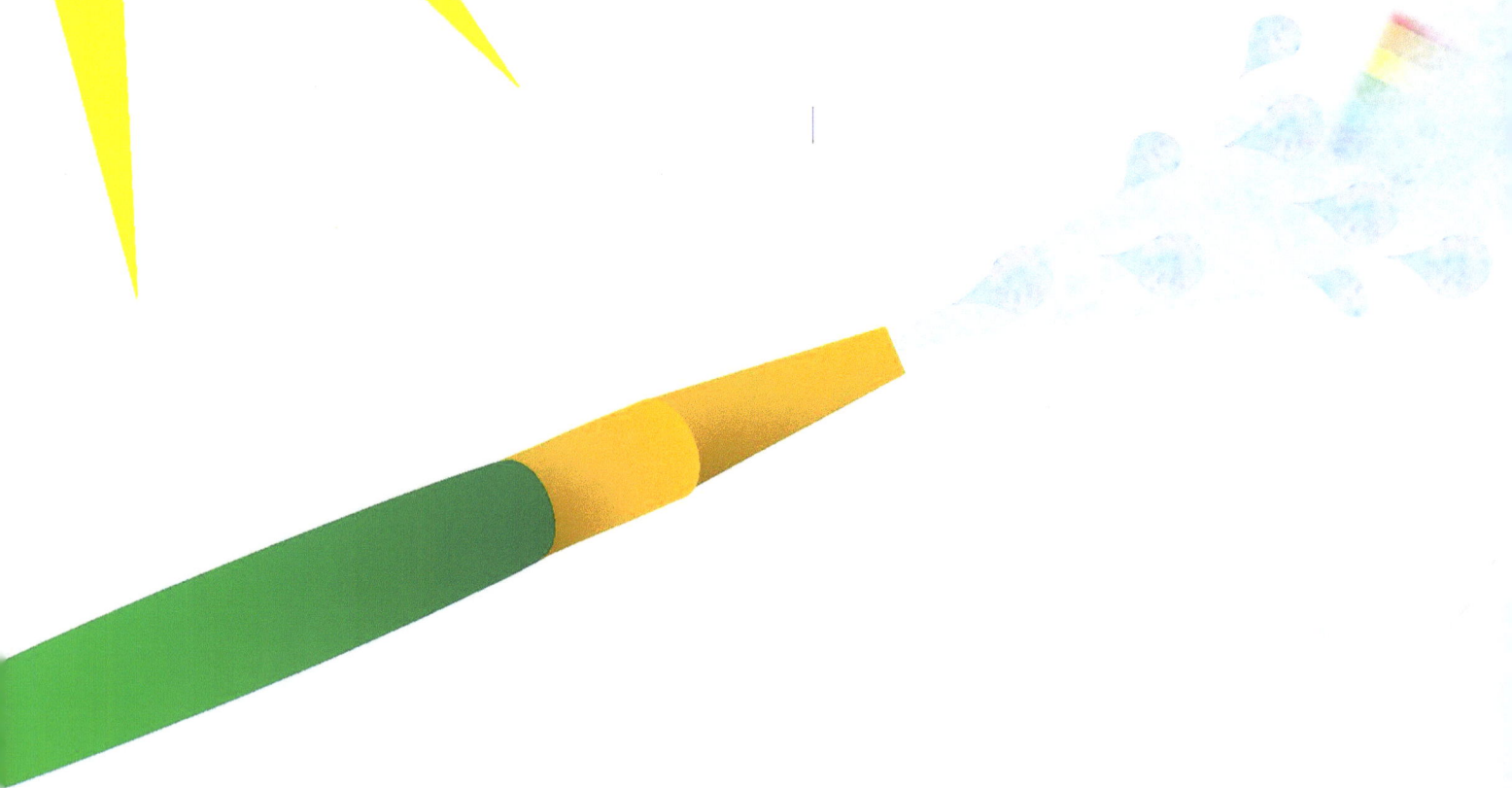

Would you like to make a rainbow a new way? We will need a small mirror, a glass of water, a piece of white paper and a flashlight.

I think the most amazing fact about rainbows is that they teach us that white light is made of seven colors. It is fascinating to me that the rainbow has had so many different meanings over so many years. A rainbow really is a wheel of fortune.

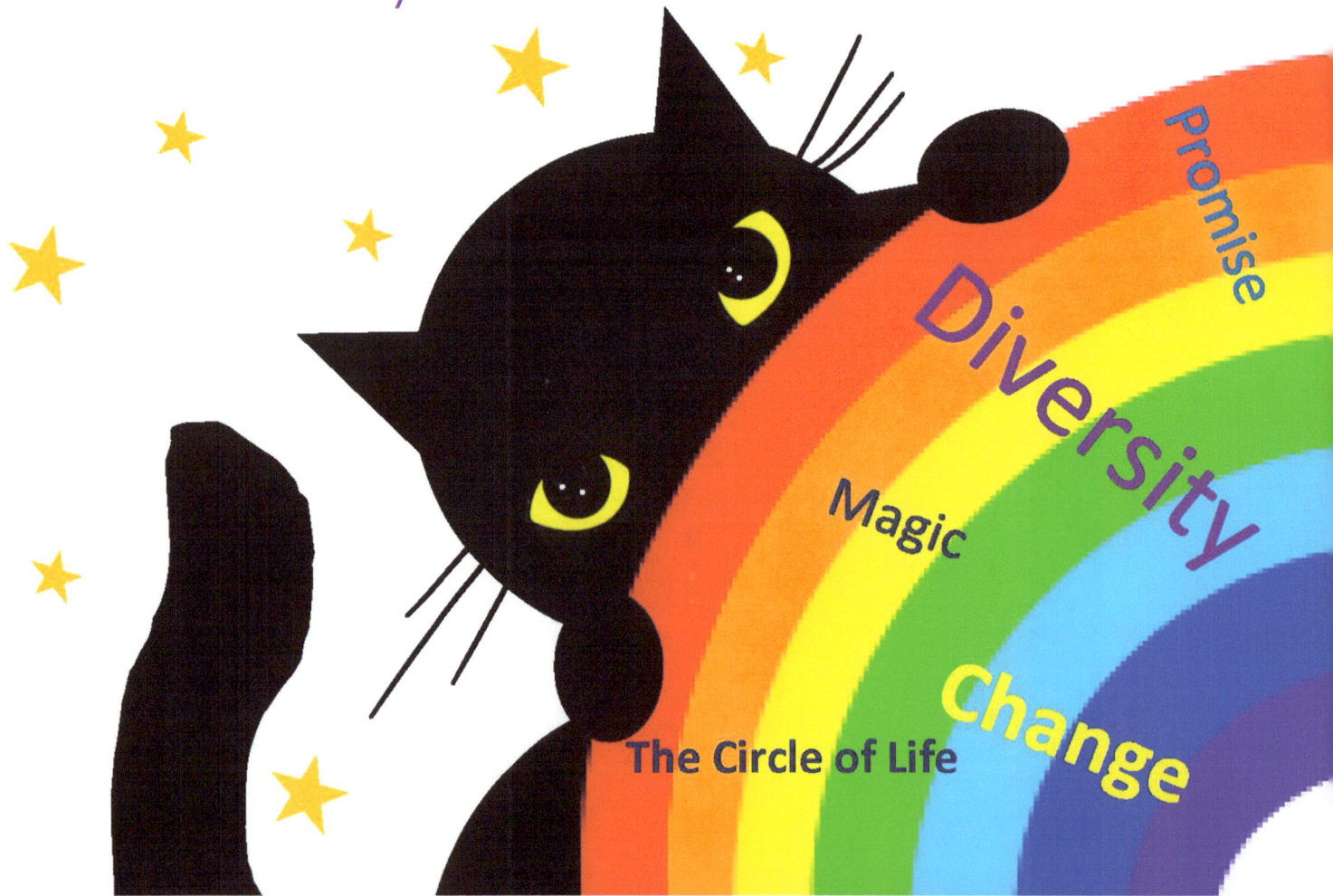

Promise

Diversity

Magic

Change

The Circle of Life

Love

Joy

Peace

Eternal Life

Acceptance

Creativity

Gratitude

Harmony

Equity

Communication

Spirituality

Good Luck

Hope

Creation

In mist, fog, rain or dew,
with sunlight shining through,
A rainbow awaits lucky you!

I really enjoyed sharing, *All About Rainbows*, with you. I have something to tell you before I take my catnap.

"Seven rainbow letters make us friends"

FRIENDS

Love Cleo

A Note from the Author

I believe that learning elevates our level of humanity and increases the quality of our relationships with each other and the world around us.

Activities

- When the weather conditions are right, look for rainbows as a family.
- Try out the experiments in this book and have your children demonstrate them to their friends.
- Use the many meanings of the **rainbow** as a guide to a great life.
- Make learning, color, and friendship an important part of your **journey**.

*Snow-White and the Seven Dwarfs was originally written by Brothers Grimm in 1812

www.ingramcontent.com/pod-product-compliance
Lightning Source LLC
Chambersburg PA
CBHW061143030426
42335CB00002B/88

9 781959 707073